YOUR KNOWLEDGE HAS VALUE

- We will publish your bachelor's and master's thesis, essays and papers

- Your own eBook and book - sold worldwide in all relevant shops

- Earn money with each sale

Upload your text at www.GRIN.com and publish for free

Bibliographic information published by the German National Library:

The German National Library lists this publication in the National Bibliography; detailed bibliographic data are available on the Internet at http://dnb.dnb.de .

This book is copyright material and must not be copied, reproduced, transferred, distributed, leased, licensed or publicly performed or used in any way except as specifically permitted in writing by the publishers, as allowed under the terms and conditions under which it was purchased or as strictly permitted by applicable copyright law. Any unauthorized distribution or use of this text may be a direct infringement of the author s and publisher s rights and those responsible may be liable in law accordingly.

Imprint:

Copyright © 2014 GRIN Verlag, Open Publishing GmbH
Print and binding: Books on Demand GmbH, Norderstedt Germany
ISBN: 978-3-668-17941-7

This book at GRIN:

http://www.grin.com/en/e-book/318609/an-inverse-approach-for-the-determination-of-a-viscous-damping-model-of

Subhajit Mondal, Sushanta Chakraborty

An Inverse Approach for the Determination of a Viscous Damping Model of Fibre Reinforced Plastic Beams using Finite Element Model Updating

GRIN Publishing

GRIN - Your knowledge has value

Since its foundation in 1998, GRIN has specialized in publishing academic texts by students, college teachers and other academics as e-book and printed book. The website www.grin.com is an ideal platform for presenting term papers, final papers, scientific essays, dissertations and specialist books.

Visit us on the internet:

http://www.grin.com/

http://www.facebook.com/grincom

http://www.twitter.com/grin_com

An Inverse Approach for the Determination of Viscous Damping Model of Fibre Reinforced Plastic Beams using Finite Element Model Updating

Subhajit Mondal and Sushanta Chakraborty
Department of Civil Engineering, Indian Institute of Technology Kharagpur, Kharagpur- 721 302, India

(Received 3 March 2014; accepted 9 April 2015)

Investigations have been carried out both numerically and experimentally to settle with a practically feasible set of proportional viscous damping parameters for the accurate prediction of responses of fibre reinforced plastic beams over a chosen frequency range of interest. The methodology needs accurate experimental modal testing, an adequately converged finite element model, a rational basis for correct correlations between these two models, and finally, updating of the finite element model by estimating a pair of global viscous damping coefficients using a gradient-based inverse sensitivity algorithm. The present approach emphasises that the successful estimate of the damping matrix is related to a-priori estimation of material properties, as well. The responses are somewhat accurately predicted using these updated damping parameters over a large frequency range. In the case of plates, determination of in-plane stiffness parameters becomes easier, whereas for beam specimens, transverse material properties play a comparatively greater role and need to be determined. Moreover, for damping matrix parameter estimation, frequency response functions need to be used instead of frequencies and mode shapes. The proposed method of damping matrix identification is able to reproduce frequency response functions accurately even outside the frequency ranges used for identification.

CONTENT

1. Introduction ... 2

2. Mathematical Formulation ... 3

3. Numerically Simulated Example ... 5

4. Experimental Investigation ... 6

5. Conclusions ... 8

References ... 9

1. INTRODUCTION

The accurate determination of dynamical responses is very important from the viewpoint of safety, serviceability, and operation of any structure. The geometrical complexities, material property distributions, existing boundary conditions, and applied loading are the key factors that influence the dynamic responses. The elastic and inertial properties are somewhat correctly represented through finite element modelling with suitable simplifying assumptions, whereas uncertainties in response prediction still remain due to imperfect boundary conditions and the presence of damping, which are difficult to deal with. No generally acceptable modelling techniques for damping have been proposed in previous research that can be confidently used for complicated structures. The damping mechanism may comprise three effects — material damping resulting from micro-structural behaviour, friction damping resulting from looseness at boundaries, and environmental damping effects, such as interaction with the surrounding fluid. Depending on the practical situation, one or more component may be less significant than the others, making the modelling effort easier for the particular structure under consideration.

Although the phenomenon of damping is mostly nonlinear, the assumption of small damping makes many equivalent linear models practically acceptable. For example, Dowell and Schwartz[1] presented a methodology for accounting for dry friction damping arising from axial sliding of surfaces inside supports, and concluded from studies on plates and beams that an equivalent linear viscous damping ratio can be agreed upon, even if nonlinear Coulomb law for the friction and geometric nonlinearity of beams are present. Tang and Dowell[2] further investigated experimentally to verify the theory presented in the previous paper. It was concluded that the methodology works well in lower mode ranges, especially with the fundamental mode. Sometimes, it will be possible to accept on practical terms the linear damping models for much more complicated environmental effects, such as interaction of a beam with the surrounding air, etc. Filipiak, et al.[3] presented such an approach to determine the effects of air damping on small beams housing miniature sensors. How far such efforts are applicable to realistic full scale structures remain an open question.

If linear damping is agreed upon with small damping assumptions, it can be modelled as a multiplier of conveniently chosen state variables with constant coefficients. The success of such a model can be judged by its ability to replicate the actual observable responses over a frequency range of interests. Then, the entire domain of linear modal testing can be employed and a damping matrix can finally be put forward in the model to be treated in a fashion similar to the stiffness and mass matrices. Mostly, instantaneous velocity is chosen as the state variable, and the damping can be stated to be viscous.[4]

Fibre Reinforced Plastics (FRP) have long been used in weight-sensitive aerospace, naval, automotive, and high performance sports applications. However, it took some time for the engineering community to appreciate the other positive aspects of FRPs, such as durability, fatigue, and corrosion resistance to pave the way for its infrastructural applications. Recently, many standard structural forms such as various beam sections, plates, and shells have been routinely manufactured and employed for structural applications. Pultruded sections in regular forms such as rectangular, angle, 'T', etc., are likely to replace most of the current infrastructures made of conventional material such as steel. Condition assessment and health monitoring of such huge infrastructures made of FRPs will be a challenge in the future, especially if they degrade over long periods of time, but still remain serviceable. The existing stiffness and damping properties need to be correctly assessed from time to time using a reliable non-destructive inverse technique. Unlike the stiffness and inertia parameters, uncertainties in damping parameters will further increase, as the mechanism may include one or more effects which were initially absent. For example, loosening of joints and supports may result in increased friction component of damping as time passes.

Literature related to the modelling of damped FRP structures is very scant. Zhuang and Crocker[5] presented a review on the damping of composite structures. Gelman, et al.[6] proposed a methodology of diagonalisation of the damping matrix based on measured frequency response functions (FRF). Akrout, et al.[7] conducted numerically simulated investigations of vibroacoustic behaviour of two thin film-laminated glass panels in the presence of a fluid layer. Assaf[8] analysed sandwich composite beams and investigated the effects of ply-stacking sequences, core-to-face stiffness ratio, etc. on natural frequencies and modal damping.

The current literature is very rich in information related to inverse detection of stiffness parameters from measured vibration responses,[9] whereas only limited attempts have been made to inversely estimate the damping parameters for FRP type of structures. The main reason is that for small damping, the resonant frequencies and mode shapes change very little with damping coefficients, but the responses change drastically, especially near resonances. Literature related to damping identification using a beam type of specimen is very rare. Reix, et al.[10] used the FRF information of a beam to update the damping matrix using a nonlinear least square optimization technique.

Inverse detection of damping using an iterative procedure demands proper forward simulation of the damped responses of the FRP structures in the iterative loop. The most popular forward damping model is due to Rayleigh, in which the damping matrix is assumed as a weighted linear combination of the mass and stiffness matrix

$$C = a_0 M + a_1 K. \quad (1)$$

A more generalised viscous proportional damping matrix has been proposed by Caughey and Kelly,[11] and can be written as

$$C = M \sum_{n=0}^{r-1} a_n \left[M^{-1} K \right]^n. \quad (2)$$

Woodhouse[12] has given an account of various linear damping models useful for structural applications. The main difficulty of all such models is that the damping parameters remain somewhat insensitive to frequency variations. Moreover, stiffness and mass distributions should be exactly determined a-priori, which is impossible in most practical cases. Adhikari[13] incorporated the frequency variation of damping factors within the framework of a generalised damping model. As a continuation of the above methodology, Adhikari and Phani[14] proposed a proportional damping matrix obtained from a single driving point FRF. Minas and Inman[15] used incomplete experimental modal data and reduced mass and stiffness matrices to identify a non-proportional damping matrix in a weighted least square sense. Lancaster and Prells[16] used the inverse spectral method to estimate the damping matrix from complex eigenvectors. Pilkey[17] proposed direct and iterative approaches for damping matrix reconstruction. Friswell, et al.[18] used a direct approach to identify damping and the stiffness matrix together using FRF information. Chen and Tsuei[19] distinguished between the viscous and structural damping components from the measured complex FRF matrix. Some investigators tried to estimate mass, stiffness, and damping matrices together.[20]

The main drawback of the investigations on damping matrix identification proposed in a great deal of research is that they demand availability of accurate information about the stiffness and mass of the system, as well as the availability of accurate modal properties. Even in newly built FRP structures, there are large variations between the predicted stiffness parameters as compared to those existing, due to the fact that the structural fabrication and material fabrication are one unified process for FRP, and the actual existing material properties vary greatly from those mentioned in manufacturer's manual or in standard handbooks. For the FRP type of anisotropic layered composites, such uncertainties are greater as compared to similar constructions made of isotropic and homogeneous materials. It thus appears that a proper inverse regularised technique augmented by a-priori stiffness estimation procedure will be appropriate for realistic damping parameter identification. If the damping is small, which is the case in most practical structures, a linear model will suffice.

The objective of the present investigation is to apply a gradient-based model updating technique to estimate viscous damping parameters along with the stiffness parameters for pultruded FRP beams using measured FRFs. The efficiency of the algorithm will be judged by comparing the regenerated FRFs to the experimentally obtained values to examine if the responses match accurately. Information related to FRF-based updating is abundant in current literature,[21] although most of it is related to the estimation of stiffness parameters as far as applications to FRP structures are concerned.

The complete process for identification of damping of FRP beams includes a-priori estimate of stiffness parameters using measured modal and FRF data, converged finite element modelling, correlations between them, and finally updating the global proportional damping parameters using the gradient-based inverse sensitivity technique in a nonlinear least square sense.[22] The methodology is first established through a numerically simulated example, followed by real experimental case studies involving different boundary conditions.

2. MATHEMATICAL FORMULATION

The equation of motion of a multiple-degrees-of-freedom system in a discretized form can be written as

$$M\ddot{x}(t) + C\dot{x}(t) + Kx(t) = f(t); \quad (3)$$

where M, K, and C are the mass, stiffness, and damping matrix, respectively. Here, equivalent viscous damping has been considered as the major dissipation mechanism. In modal coordinates, the equation can be written as a set of single-degree-of-freedom (SDOF) uncoupled equations

$$[m]\ddot{x}(t) + [c]\dot{x}(t) + [k]x(t) = \{u\}(t); \quad (4)$$

where

$$[m] = \phi^T[M]\phi = \text{modal mass matrix},$$
$$[k] = \phi^T[K]\phi = \text{modal stiffness matrix},$$
$$[c] = \phi^T[C]\phi = \text{modal damping matrix}. \quad (5)$$

The free vibration equation can be expressed as

$$\left(-\omega^2[m] + i\omega[c] + [k]\right)\{u\} = 0. \quad (6)$$

If damping is neglected, the undamped equation of motion can be solved from the eigenequation

$$Ku = \omega^2 Mu. \quad (7)$$

These undamped eigenvalues and eigenvectors can be used to form the acceleration FRFs, and can be expressed as a function of modal damping factor

$$H_{ij}(\omega) = -\omega^2 \sum_{k=1}^{N} \frac{\phi_{ik}\phi_{jk}}{\omega_k - \omega + 2i\omega\omega_k\xi_k}. \quad (8)$$

It is to be noted that the Rayleigh damping coefficients can be related to the modal damping factor, as shown below:

$$\xi_i = \frac{1}{2}\left(\frac{a_0}{\omega_i} + a_1\omega_i\right). \quad (9)$$

The disadvantage of Rayleigh damping is that the effects of the higher modes are usually weighted more than the lower modes. As only two modes are used at a time for the estimation of modal damping, the effects of other modes cannot be taken care of. The present investigation is focused on removing this difficulty by including the effects of multiple modes through measured FRFs within a frequency range of interest for a FRP beam.

The expression for the FRF can be modified to include the damping coefficients as

$$H_{ij}(\omega) = -\omega^2 \sum_{k=1}^{N} \frac{\phi_{ik}\phi_{jk}}{\omega_k - \omega + i\omega\left(a_1\omega^2 + a_0\right)}; \quad (10)$$

where H_{ij} is the acceleration response at point i due to excitation at point j. As it is a common practice to deploy accelerometers for measuring accelerations directly and compute displacement and velocities as derived quantities as and when required, all response quantities are expressed in terms of accelerations in this paper. In case of other measurement techniques employing measured displacements as first hand information, such as in full field measurements using scanning Laser Doppler Vibrometer types of non-contact devices, the formulations can be modified to deal with displacements directly.

Since the order of magnitude of the terms of the damping matrix is much lower as compared to the stiffness and mass matrices, it will be efficient to have the stiffness properties updated prior to the updating of the damping parameters. Thus, a two-stage model updating algorithm is implemented here. Moreover, the global stiffness properties can be updated more efficiently with the help of measured natural frequencies and mode shapes, whereas updating the damping matrix coefficients a_1 and a_0 requires the information from measured FRFs.

At present, the inertia properties are assumed to be determined accurately, as this is generally the case in practice.

The objective functions involving the measured and modelled FRFs can be written in a weighted least square sense

$$E = \sum_{i=1}^{q} w_{ii} \|H_{\exp}(\omega) - H_{nu}(\omega)\|^2; \quad (11)$$

where w_{ii} are the weights and q is the number of FRFs considered. The sensitivities of these FRFs with respect to the damping or elastic parameters can be computed using

$$S_{ij} = \left[\frac{\partial H(\omega)}{\partial r_j}\right]; \quad (12)$$

where $i = 1$ to n and $j = 1$ to m. Here, the order of the sensitivity matrix is $n \times m$. The linearized first order approximation of the relationship between changes in measured modal properties (i.e. frequencies, mode shapes or FRFs) and the changes in the parameters to be updated can be related through the first order sensitivity matrix

$$\{\Delta f\} = [S]\{\Delta r\}. \quad (13)$$

In the updating process, changes (Δr) are made to the initial guesses of parameters within reasonable bounds, and the finite element model of the pultruded FRP beam is updated as follows:

$$\{r\}_{i+1} = \{r\}_i + \{\Delta r\}_i. \quad (14)$$

The error between the experimental observation and the finite element modelling is thus minimised through this inverse sensitivity method. In the present investigation, the parameters can be the in-plane elastic material constants, such as the Young's modulus and the shear modulus, the out-of-plane shear modulus, and the modal damping coefficients. A Block Lanczos Algorithm has been implemented for the eigensolutions.

In an inverse problem related to anisotropic materials, mode sequences need to be properly checked using certain established correlation criteria. In the present investigation, Modal Assurance Criteria (MAC) is used at each step of iteration to determine how similar or dissimilar the analytical modal vector is as compared to the experimentally measured modes; a value close to 1 indicates good correlations:[23]

$$MAC(\phi_{nu}, \phi_{\exp}) = \frac{\left|\{\phi_{nu}\}^T\{\phi_{\exp}\}\right|^2}{\left(\{\phi_{nu}\}^T\{\phi_{\exp}\}\right)\left(\{\phi_{\exp}\}^T\{\phi_{\exp}\}\right)}. \quad (15)$$

Here, ϕ represents the realised eigenvectors from the measured complex modes. The analytical and experimental FRFs are similarly correlated using Signature Assurance Criteria (SAC),[22] which is basically a global Frequency Response Assurance Criteria (FRAC):[24]

$$SAC(H_{nu_i}, H_{\exp_i}) = \frac{\left(\left|H_{\exp_i}^T\right|\left|H_{nu_i}\right|\right)^2}{\left(\left|H_{\exp_i}^T\right|\left|H_{\exp_i}\right|\right)\left(\left|H_{nu_i}^T\right|\left|H_{nu_i}\right|\right)}. \quad (16)$$

Furthermore, Cross Signature Assurance Criteria (CSAC) is the correlation function, checking the FRF correlations[22] lo-

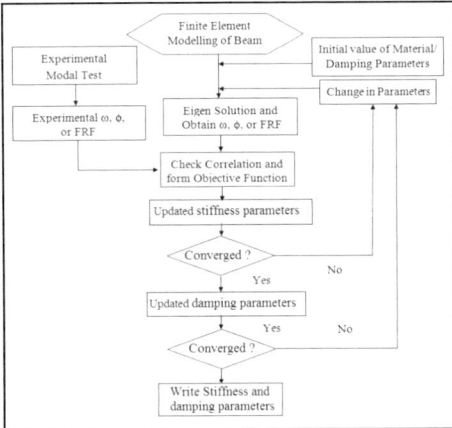

Figure 1. Flow chart of the model updating algorithm.

cally:

$$CSAC(\omega_k) = \frac{\left|H_{\exp_i}^T(\omega_k)H_{\mathrm{nu}_i}(\omega_k)\right|^2}{\left(H_{\exp_i}^T(\omega_k)H_{\exp_i}(\omega_k)\right)\left(H_{\mathrm{nu}_i}^T(\omega_k)H_{\mathrm{nu}_i}(\omega_k)\right)},$$
$$k = 1, 2, \ldots, N. \quad (17)$$

The amplitude correlations of FRFs are taken care of by Cross Signature Scale Factor (CSF):[22]

$$CSF(\omega_k) = \frac{2\left|H_{\exp_i}^T(\omega_k)H_{\mathrm{nu}_i}(\omega_k)\right|}{\left(H_{\exp_i}^T(\omega_k)H_{\exp_i}(\omega_k)\right) + \left(H_{\mathrm{nu}_i}^T(\omega_k)H_{\mathrm{nu}_i}(\omega_k)\right)},$$
$$k = 1, 2, \ldots, N. \quad (18)$$

First, an isotropic beam is investigated numerically to see if the damping parameters can be conveniently determined. The 'experimental' FRFs in this simulated study are also determined using the same finite element programming. Subsequently, a real experiment is conducted on a Pultruded FRP beam, both in cantilever and under free boundary conditions. For the updating process, the initial finite element model computes the FRFs using the undamped frequencies, modal vectors, and initially assumed modal damping factors. Analytical and experimental FRFs are correlated as explained earlier to form the objective functions. A first order sensitivity matrix is computed for the selected parameters by a finite difference approximation of variables. Finally, the inverse sensitivity method is used to update the stiffness parameters, first using the modal information, followed by an estimation of modal damping coefficients using the FRFs. A Bayesian approach is used to include the variance of the response data. The entire procedure is explained through a flow chart in Fig. 1.

3. NUMERICALLY SIMULATED EXAMPLE

To check the stability and efficiency of the algorithm described above, first a numerically simulated example involving a rectangular isotropic beam of dimensions 500 mm × 40 mm and having thickness of 10 mm is considered and is shown in Fig. 2. A three-noded quadratic beam element (B32)[25] is used

Table 1. Numerically simulated 'experimental' frequencies and assumed modal damping factors.

Mode	Frequency (Hz)	Modal damping factor (%)
1	24.93	5.0E-2
2	99.28	1.3E-2
3	155.90	1.1E-2
4	435.05	1.25E-2
5	528.57	1.50E-2
6	604.20	0.8E-2

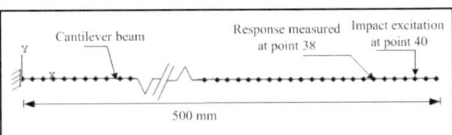

Figure 2. Numerical model of the cantilever beam.

for the finite element modelling of the isotropic beam.[25] The present investigation deals with the average stiffness properties and global average damping parameters. A 40-mesh division along the length was found to be sufficient for convergence of eigenproperties. The actual existing Young's modulus E and in-plane Poisson's ratio ν are taken to be 30 GPa and 0.3, respectively. The mass density is assumed as 2012 kgm^{-3}. With the above set of data, the simulated 'experimental' modal properties are computed and presented in Table 1, along with the assumed modal damping factors.

As explained earlier, the determination of material constants from the modal data using the inverse sensitivity method is taken up first. It has been observed that changing the modal damping coefficients to have a different set of 'experimental' modal data has very little effect on the accuracy of the estimation of these stiffness parameters, and thus is not reported here. The updated values of the elastic material constants were used for further updating of the damping coefficients, which requires the simulated 'experimental' FRF data.

It is readily observed that the estimated values of the damping parameters a_0 and a_1 differ depending upon the modes considered. The results are shown in Table 2 for a few selected arbitrary combinations of modes using Eq. (9).[26] The corresponding values of ω_i and ξ_i are taken from Table 1. The first three sets show the variations of the two estimated damping parameters due to the incorporation of up to the first five modes.

While implementing the inverse FRF-based updating algorithm, these values of a_0 and a_1 are chosen as initial values to see if all trials converge to a unique set of parameters. To check the robustness of this FRF-based inverse algorithm, two additional arbitrary sets of values of a_0 and a_1 are also chosen (Trial_4 and Trial_5) that do not immediately correspond to any combination of modes and may not have any physical significance.

Figure 3 shows the monotonic convergence curves for both the parameters, the final values of which are $a_0 = 14.95$ rad/s and $a_1 = 31.60$E-6 s/rad, respectively. The updated mass proportional damping coefficient converged to a value that is quite higher, while the stiffness proportional damping coefficient converged to a somewhat lower value.

A typical set of regenerated FRF curves with different mode combinations are shown in Fig. 4. It clearly shows that the most accurate global representation of damping parameters de-

Table 2. Initial values of damping parameters for numerically simulated example.

	Mode considered for average values of ω_1 and ξ_1	Mode considered for average values of ω_2 and ξ_2	a_0	a_1
Trial_1	1	2	2.48	9.55E-6
Trial_2	1, 2	5	3.74	4.33E-5
Trial_3	2	3	2.00	5.87E-5
Trial_4	—	—	40	5.0E-6
Trial_5	—	—	50	1.0E-6

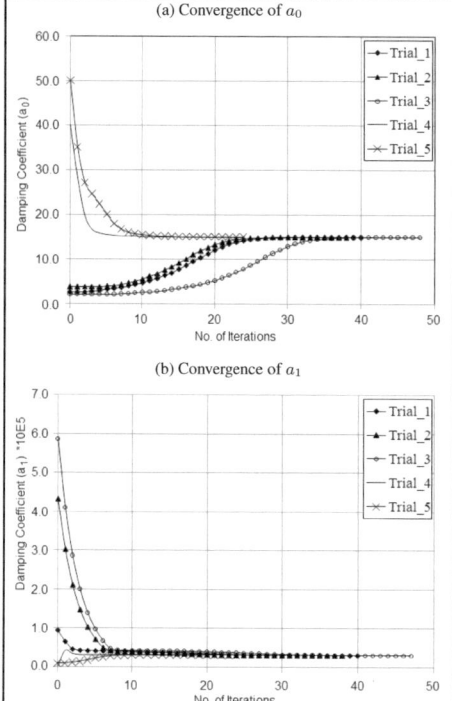

Figure 3. Convergence curves for damping coefficients.

pends upon the participation of modes. If that matches with the selected modes, then only the SAC value indicating better correlation between the observation and model will approach unity. This is difficult to predict in practice, and several trials with various combinations of modes are necessary before settling with the most appropriate solution.

4. EXPERIMENTAL INVESTIGATION

Equipped with the knowledge gained from the numerically simulated example, a rectangular FRP composite beam of the same size is fabricated using the pultrusion process with Woven Roving (WR) glass fibres and an epoxy matrix. The exact final average thickness of the beam comes to be 10.12 mm. The mass density was measured to be exactly 2012 kgm^{-3}. First, the modal testing was carried out with a fixed boundary at one end of the beam, the other end being free. An

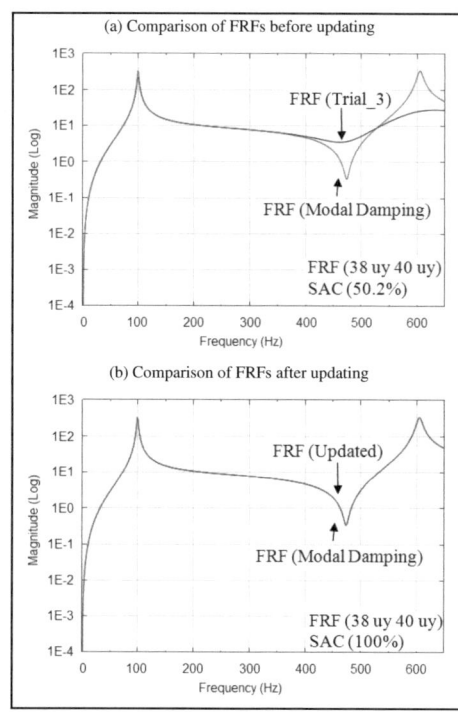

Figure 4. Comparison of FRFs computed using modal damping to those computed using trial values and updated values of damping parameters.

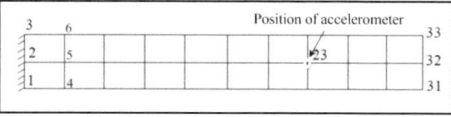

Figure 5. Measurement grid points of the FRP cantilever beam.

impact hammer fitted with a force transducer (B&K, number 8206-002) was used for exciting the beam at different predefined locations, and the resulting responses were picked up by an accelerometer (IEPE DeltaTron 4507) at a particular node. Both the signals were Fourier transformed in a B&K spectrum analyser, and the FRFs were obtained utilising the PULSE-LabShop software.[27] The frequencies, modes shapes, and modal damping factors were extracted from the measured FRFs using the post-processing software MEScope.[28] Figure 5 shows the position of the accelerometer (point 23) and also the nodes where forces were imparted through the impact hammer in turn (33 nodal points altogether). Figure 6 shows the experimental setup for the modal testing. Very heavy steel billets are used to ensure proper fixity after using a properly bolted connection at the cantilever end. Accelerometers were placed near the supports to check whether near zero support mobility is achieved during testing.

The finite element modelling of the beam was done with shell element (S8R),[25] implementing an equivalent single layer

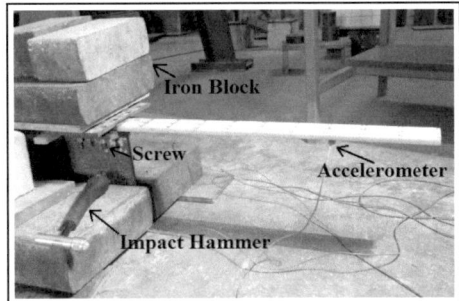

Figure 6. Experimental modal testing of the FRP cantilever beam.

Table 3. Updated and experimentally observed natural frequencies and damping factors of the FRP cantilever beam.

Mode No.	Updated Freq. (Hz)	Exp. Freq. (Hz)	Modal Damping Factors (%)
1	25.63	25.61	15.56
2	160.06	160.85	2.84
3	399.28	399.28	2.43
4	445.54	442.65	1.47
5	865.86	867.96	1.26

Table 4. Experimentally obtained and updated material parameters.

Parameters	Experimentally obtained elastic parameters (GPa)	Updated and finally used elastic parameters (GPa) for damping matrix updating
E_x	33.05	31.66
E_y	31.80	31.80
G_{xy}	5.73	6.30
G_{xz}	—	5.37
G_{yz}	—	5.37

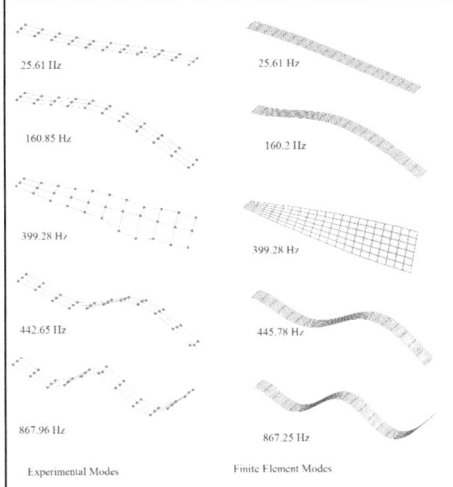

Figure 7. Experimentally and numerically obtained mode shapes for the FRP cantilever beam.

Table 5. Initial values of damping coefficients from experimentally obtained modal damping factors of the FRP cantilever beam.

	Mode considered for average values of ω_1 and ξ_1	Mode considered for average values of ω_2 and ξ_2	a_0	a_1
Trial_1	1	2	7.93	4.71E-5
Trial_2	1, 2	5	17.07	6.42E-6
Trial_3	2	3	7.22	8.24E-5
Trial_4	2	4	8.78	2.19E-5
Trial_5	—	—	60	1.00E-4
Trial_6	—	—	20	2.00E-6

theory for layered composites.[29] The parameters selected for updating are the in plane stiffness parameters of the beam, out-of-plane shear modulus, and the damping constants. First, the material constants are updated, followed by the estimation of damping coefficients. The initial values to start the iterative model updating process are selected from standard handbooks and manufacturer's data. Figure 7 shows the comparison between properly correlated experimental and numerical mode shapes.

Frequencies obtained using updated elastic material parameters, along with the experimentally measured frequencies, are shown in Table 3. The experimentally obtained modal damping factors are also shown.

The final MAC values between the experimental and updated mode shapes indicate an excellent correlation. To have better insight into the global updating process, the correlation quantities CSAC and CSF were also computed, and excellent correlations were achieved, except near some anti-resonant points.

Next, static characterisation tests were carried out using coupons that were prepared and tested quasi-statically as per ASTM standard (No.D3039/D3039M),[30] and the results are presented in Table 4.

The experimentally obtained Poisson's ratio is 0.15. For the updating of the damping parameters, the initial values are selected again from Eq. (9), in the same way as they were selected in the numerically simulated example. A few such selected sets of damping coefficients are shown in Table 5, along with two arbitrary values to test the robustness of the algorithm from distant points in this practical example.

The FRFs using different trial values of damping parameters are shown in Fig. 8, and it is clear that they still differ from the experimentally observed values. The regenerated responses using the updated damping parameters, however, match almost exactly with the experimentally obtained FRFs, as shown in the Fig. 9. The SAC value also approaches 1, indicating very good global correlations.

Figure 10 shows comparisons between the experimental FRFs and the numerically regenerated FRFs that use the experimentally obtained modal damping factors. The convergence curves of the damping parameters from various initial values are shown in Fig. 11, and are found to be monotonic in all cases. The last updated parameters are found to be $a_0 = 42.25$ rad/s and $a_1 = 1.10$E-5 s/rad, respectively, for this FRP beam, as shown in the Fig. 11, considering five modes altogether. The apparent improvement in global response prediction can be attributed to the incorporation of a number of modes, rather than using only a few selected modes (Eq. (9)).

To test the authenticity of the methodology developed, some more FRFs are compared that were not used in the updating process, and the correlations are found to be excellent. This was observed at most of the anti-resonant points, as well

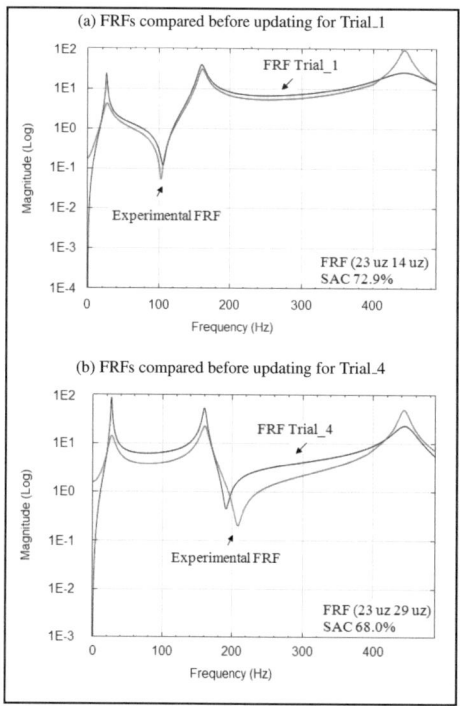
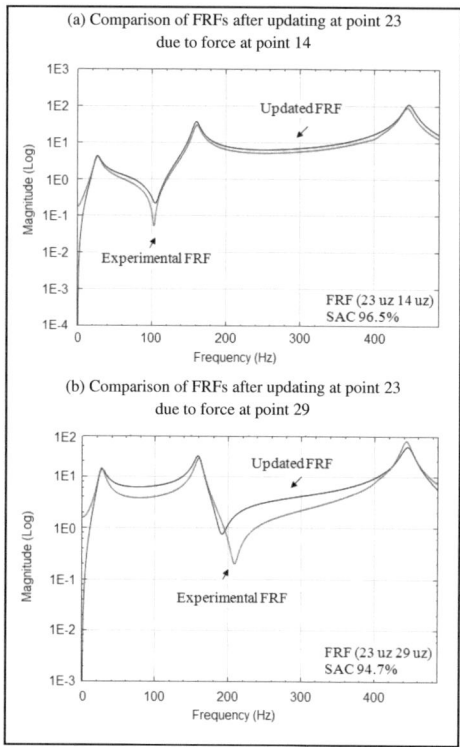

Figure 8. Typical comparison of experimental and trial FRFs at two selected points using different values of viscous damping coefficients.

Figure 9. Typical comparison of experimental and updated FRFs at two selected points.

(as shown in Fig. 12). Even, the modes beyond the frequency ranges considered also showed improved correlations (as shown in Fig. 13).

The experimental investigation was then extended to cater to the free boundary conditions, as well, and a two-step finite element model updating procedure was performed. The free boundary conditions were achieved by hanging the beam from soft rubber threads of sufficient length so that the frequency of oscillation was much lower as compared to the fundamental frequency of the free-free beam. The last measured width and length of the beam were found to be 40 mm and 501 mm, respectively, whereas the average thickness was measured to be 10.12 mm. The schematic diagram showing the measurement points, as well as a photograph of modal testing under free boundary conditions, are shown in Fig. 14.

First the material properties were updated, followed by updating the viscous damping parameters from the FRF data using the sensitivity-based model updating algorithm.

The updated in-plane Young's moduli (E_x and E_y) were found to be 31.85 GPa and 30.86 GPa, respectively. The in-plane shear modulus (G_{xy}) was updated to a value of 6.31 GPa, and the out-of-plane shear modulus (G_{xz} and G_{yz}) were updated to 4.77 GPa and 5.56 GPa, respectively. The Poisson's ratio was measured to be 0.15. The experimentally obtained first four frequencies were 165.87 Hz, 453.39 Hz, 800.44 Hz, and 886.53 Hz, and the modal damping coefficients were measured to be 2.59%, 1.30%, 2.52%, and 1.40%, respectively. The experimental mode shapes are compared in Fig. 15 to the finite element mode shapes, indicating very good correlations.

A typical set of FRF curves before and after updating are also shown in Fig. 16, indicating good correlations in terms of improved SAC values. The updated damping parameters are found to be 49.3 rad/s and 5.09E-6 s/rad, which reproduces the modal damping almost exactly.

5. CONCLUSIONS

A finite element model updating algorithm using measured frequency response functions has been implemented to estimate proportional damping parameters of a fibre-reinforced plastic beam with different boundary conditions over a selected frequency range of interest. It has been observed that the material constants need to be updated a-priori before estimating the damping parameters. The number of frequencies to be included is case-specific, and for this example it gives very good accuracy with only a few modes, with which even the out-of-range frequency responses were regenerated with acceptable accuracy. At present, the methodology assumes equivalent viscous damping for all combined effects, such as boundary friction, etc.

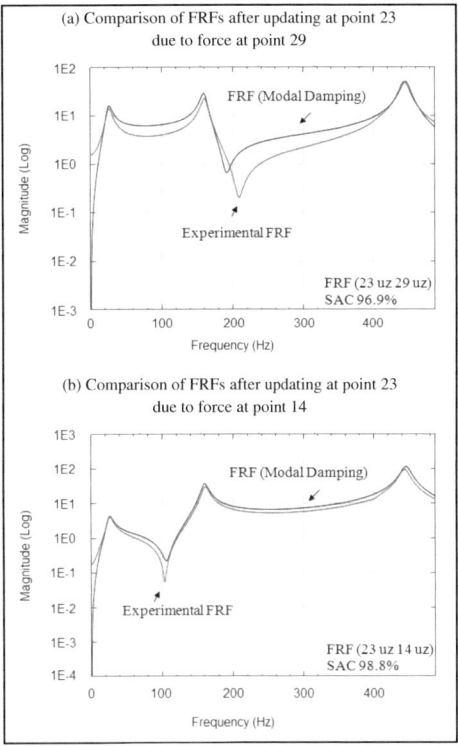

Figure 10. Typical comparison of experimental and numerically regenerated (using the modal damping factors) FRFs.

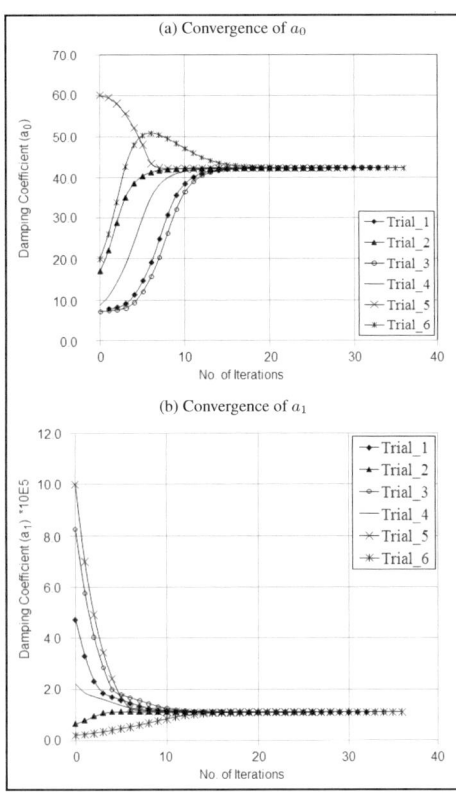

Figure 11. Convergence Curve for damping parameters of the FRP cantilever beam.

REFERENCES

[1] Dowell, E. H. and Schwartz, H. B. Forced response of a cantilever beam with a dry friction damper attached, Part-1: Theory, *Journal of Sound and Vibration*, **91** (2), 255–267, (1983). http://dx.doi.org/10.1016/0022-460x(83)90901-x.

[2] Tang, D. M. and Dowell, E. H. Damping in beams and plates due to slipping the support boundaries, part 2: Numerical and experimental study, *Journal of Sound and Vibration*, **108** (3), 509–522, (1986). http://dx.doi.org/http://dx.doi.org/10.1016/s0022-460x(86)80044-x.

[3] Filipiak, J., Solarz, L., and Zubko, K. Analysis of damping effect on beam vibration, *Molecular and Quantum Acoustics*, **27**, 79–88, (2006).

[4] Rayleigh, L. Theory of sound (Two Volumes), Dover Publications, (1877).

[5] Zhuang, Li and Crocker, M. J. A review on vibration damping in sandwich composite structures, *International Journal of Acoustics and Vibration*, **10** (4), 159–169, (2005).

[6] Gelman, L., Jenkin, P., and Petrunin, I. Vibro-acoustical damping diagnostics: Complex frequency response function versus its magnitude, *International Journal of Acoustics and Vibration*, **11** (3), 120–124, (2006).

[7] Akrout, A., Hammami, L., Chafik, K., Ben Tahar, M., and Haddar, M. Vibroacoustic damping simulation of two laminated glass panels coupled to viscothermal fluid layer, *International Journal of Acoustics and Vibration*, **15** (2), 79–90, (2010).

[8] Assaf, S. Finite element vibration analysis of damped composite sandwich beams, *International Journal of Acoustics and Vibration*, **16** (4), 163–172, (2011).

[9] Mottershead, J. E. and Friswell, M. I. Model updating in structural dynamics: A survey, *Journal of Sound and Vibration*, **167** (2), 347–375, (1993). http://dx.doi.org/10.1006/jsvi.1993.1340.

[10] Reix, C., Tombini, C., Gerard, A., and Dascotte, E. Updating the damping matrix using frequency response data, *Proc. 14th International Modal Analysis Conference*, (1996).

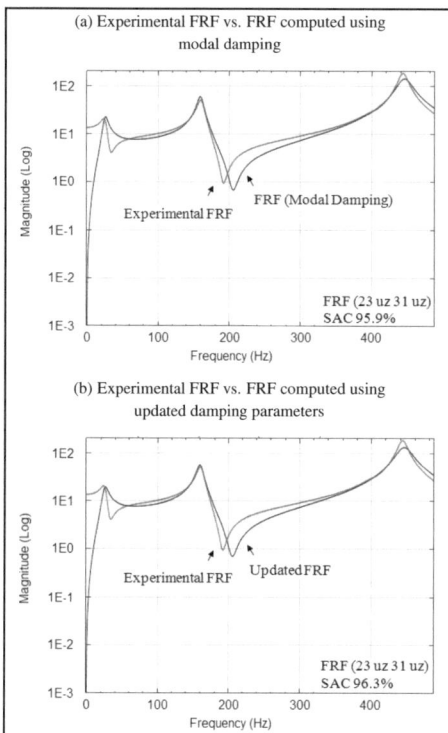

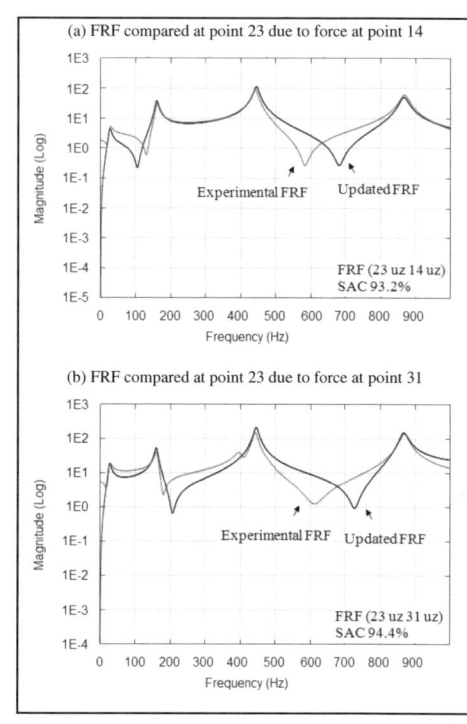

Figure 12. Typical comparison of experimental FRFs to those computed using modal damping and using updated damping parameters.

Figure 13. Typical comparison of experimental FRFs to updated FRFs at different points at higher frequency ranges.

[11] Caughey, T. K. Classical normal modes in damped linear dynamic systems, *Journal of Applied Mechanics*, **32** (3), 583–588, (1960). http://dx.doi.org/10.1115/1.3643949.

[12] Woodhouse, J. Linear damping models for structural vibration, *Journal of Sound and Vibration*, **215** (3), 547–569, (1998). http://dx.doi.org/10.1006/jsvi.1998.1709.

[13] Adhikari, S. Damping modelling using generalized proportional damping, *Journal of Sound and Vibration*, **293** (1–2), 156–170, (2006). http://dx.doi.org/10.1016/j.jsv.2005.09.034.

[14] Adhikari, S. and Phani, A. S. Experimental identification of generalized proportional viscous damping matrix, *Journal of Vibration and Acoustics*, **131** (1), (2009). http://dx.doi.org/10.1115/1.2980400.

[15] Minas, C. and Inman, D. J. Identification of a nonproportional damping matrix from incomplete modal information, *Journal of Vibration and Acoustics*, **113** (2), 219–224, (1991). http://dx.doi.org/10.1115/1.2930172.

[16] Lancaster, P. and Prells, U. Inverse problems for damped vibrating systems, *Journal of Sound and Vibration*, **283** (3–5), 891–914, (2005). http://dx.doi.org/10.1016/j.jsv.2004.05.003.

[17] Pilkey, D. F. Computation of a damping matrix for finite element model updating, PhD Thesis, Virginia Polytechnic Institute and State University, (1998).

[18] Friswell, M. I., Inman, D. J., and Pilkey, D. F. The direct updating of damping and stiffness matrices, *AIAA Journal*, **36**, 491–493, (1998). http://dx.doi.org/10.2514/3.13851.

[19] Chen, S. Y., Ju, M. S., and Tsuei, Y. G. Estimation of mass, stiffness and damping matrices form frequency response functions, *Journal of Vibration and Acoustics*, **118** (1), 78–82, (1996). http://dx.doi.org/10.1115/1.2889638.

[20] Fritzen, C. P. Identification of mass, damping, and stiffness matrices of mechanical systems, *Journal of Vibration, Acoustics Stress and Reliability in Design*, **108** (1), 9–16, (1986). http://dx.doi.org/10.1115/1.3269310.

[21] Imregun, M., Visser, W. J., and Ewins, D. J. Finite element model updating using frequency response function data: I. Theory and initial investigation, *Mechanical Systems and Signal Processing*, **9** (2), 187–202, (1995). http://dx.doi.org/10.1006/mssp.1995.0015.

[22] FEMtools, Dynamic Design Solutions, Version 3.6.1.

[23] Allemang, R. J. The modal assurance criterion — Twenty years of use and abuse, *Sound and Vibration*, **37** (8), 14–20, (2003).

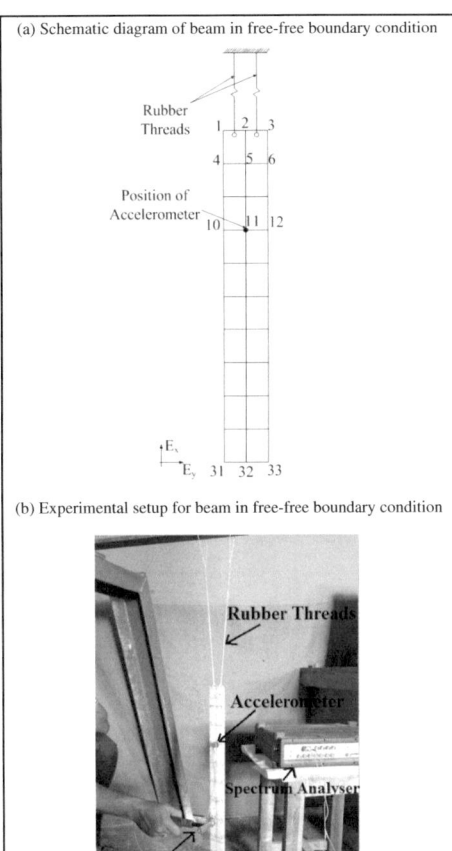

Figure 14. Experimental setup for the FRP beam with free-free boundary conditions.

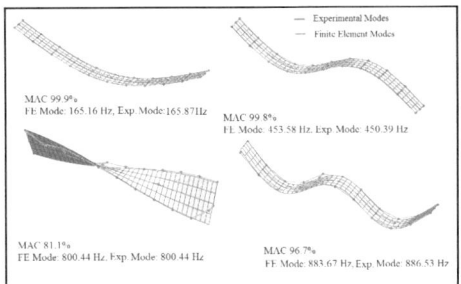

Figure 15. Comparison of experimental and finite element modes for free-free FRP beam.

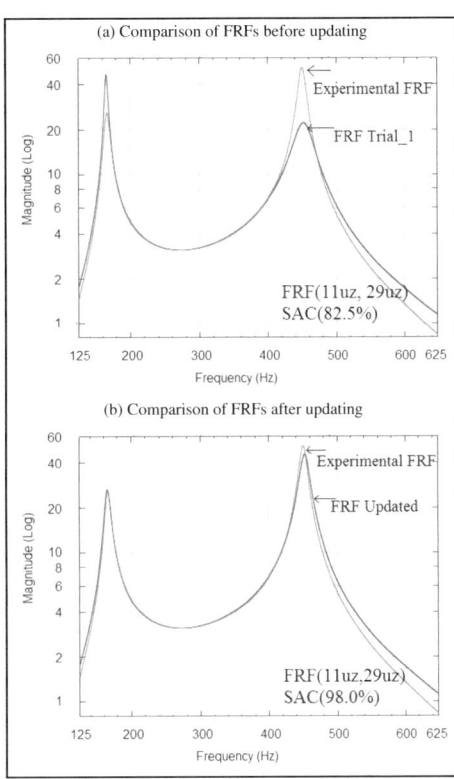

Figure 16. Typical comparison of experimental FRFs with FRFs before and after updating of damping parameters.

[24] Ewins, D. J. Modal Testing: Theory, Practice and Application, Research Studies Press Ltd., London, (2000), 442.

[25] ABAQUS, User's Manual for Version 6.10.

[26] Bathe, K. J. *Finite Element Procedures*, Prentice Hall of India, New Delhi, (2010), 796–798.

[27] PULSE LabShop, Software Package, Bruel & Kjaer, Ver. 13.1.0.246, (2008).

[28] VES ME, Vibrant Technology Inc., Ver. 4.0.0.96, (2007).

[29] Daniel, M. I. and Ishai, O. *Engineering Mechanics of Composite Materials*, Oxford University Press, (2009).

[30] Standard Test Method for Tensile Properties of Polymer Matrix Composite Materials, ASTM Standard D3039/D3039M-2008, (2008). http://dx.doi.org/10.1520/d3039_d3039m.

YOUR KNOWLEDGE HAS VALUE

- We will publish your bachelor's and master's thesis, essays and papers

- Your own eBook and book - sold worldwide in all relevant shops

- Earn money with each sale

Upload your text at www.GRIN.com and publish for free